Gourds

This book has been reviewed
for accuracy by
Jerry Doll
Professor of Agronomy
University of Wisconsin—Madison.

Library of Congress Cataloging in Publication Data

Pohl, Kathleen.
    Gourds.

    (Nature close-ups)
    Adaptation of: Hechima / Sato Yukoh.
    Summary: Describes the life cycle and growing
patterns of various kinds of gourds.
    1. Gourds—Juvenile literature.   [1. Gourds.
2. Plants]   I. Satō, Yūkō, 1928-   Hechima.
II. Title.   III. Series.
QK495.C96P64   1986   583′.46   86-26255

ISBN 0-8172-2712-1 (lib. bdg.)
ISBN 0-8172-2730-X (softcover)

This edition first published in 1987 by Raintree Publishers Inc.

Text copyright © 1987 by Raintree Publishers Inc., translated by
Jun Amano from *Gourds* copyright © 1981 by Yukoh Sato.

Photographs copyright © 1981 by Yukoh Sato.

World English translation rights for *Color Photo Books on Nature*
arranged with Kaisei-Sha through Japan Foreign-Rights Center.

1 2 3 4 5 6 7 8 9 0   90 89 88 87 86

Gourds

Adapted by
Kathleen Pohl

Raintree Publishers

Milwaukee

The seeds of a gourd.

Gourds are easy to grow. Plant seeds in the spring, when the danger of frost is past. Gourds need lots of sunshine and plenty of rainfall in order to grow well.

▶ **A young gourd plant growing.**

The roots develop many fine white hairs. These root hairs reach out in the soil to absorb water and nutrients for the growing plant. Soon, tender green seed leaves push their way up through the soil.

If you have ever seen gourds growing in a garden in autumn, you know how colorful they are. They may be red, orange, yellow, green, or multicolored. Some gourds are long and thin, others are squat and fat. Some are shaped like bottles, others like turbans. Gourds may be smooth-skinned, or covered with warts. Gourds are often grown for their unusual sizes and shapes and bright colors. They are used as decorations in people's houses.

Gourds have had many other uses throughout history. In ancient cultures, gourds were grown for food, although most were bitter-tasting. Cough syrups, ointments, and lotions were made from the fruit, or pulp, of the gourd. Lutes, drums, maracas, and other musical instruments were shaped from gourds. The gourd discussed in this book is called the loofah gourd. In recent years, it has become popular as a bath sponge.

Loofah gourds grow best in warm, tropical countries. They thrive in Mexico, Cuba, and Japan. They are also grown in the southern United States and in California.

◀ **A seed that has sprouted.**

A few days after the gourd seed is planted, it begins to sprout. The white sprout will become the root of the new plant.

◄ **A gourd plant climbing a pole.**

Gourds are climbing plants. They develop long, thin, vinelike tendrils that twine around poles and trees. As the plant climbs upward, its leaves are exposed to more sunlight.

▶ **Gourd plants growing along supporting poles.**

A few days after gourd seeds are planted in the spring, most of them will sprout and take root. Soon, tiny leaves will push their way through the soil. These are called seed leaves because they were actually formed inside the seed coat. The seed leaves contain stored nutrients that are important for the young plant's growth.

Before long, true leaves begin to grow between the seed leaves. As the nutrition in the seed leaves is used up, they drop off. The true leaves take over the task of providing food for the growing plant. In the green leaves of gourd plants, as in other plants, food is produced by a complex process called photosynthesis.

◄ **Seed leaves and true leaves.**

As the plant's true leaves develop, the nutrition in the seed leaves is used up. Soon, the seed leaves will fall off.

▶ **Insects attracted to a gourd plant.**

The loofah gourd plant has glands, called nectaries (arrows), on the back of its leaves. The nectaries secrete nectar, a sweet liquid that attracts insects. In the left photo is an aphid; in the right photo an ant is attracted to the loofah gourd's nectar.

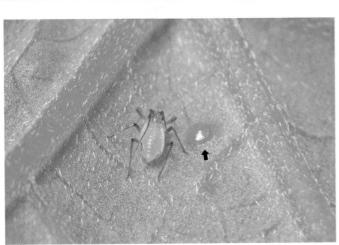

◀ **A ladybug eating a leaf from a gourd plant.**

This species of ladybug uses its sharp mandibles to mark a semicircle on the gourd leaf. Then it feeds inside the circle. It makes tiny holes, eating everything except the veins in the leaf.

▶ **A gourd leaf that has been partly eaten by a ladybug.**

There are a few kinds of insects that feed on the gourd plant. Ladybugs are generally considered to be helpful to farmers and gardeners because they eat insect pests that would otherwise destroy crops. But in Japan, there is a kind, or species, of ladybug that eats plants, rather than insects. This ladybug is attracted to the tender young leaves of the growing gourd plant. It uses its sharp jaws, called mandibles, to chew tiny holes in the plant leaves. This can cause great damage in areas where gourds are grown for commercial purposes.

Sometimes, insect pests are controlled by other insects. For instance, ants often drive ladybugs from gourd plants. It is their way of keeping the gourd plant's supply of sweet nectar for themselves.

▶ **Measuring the size of a gourd leaf.**

Find a gourd leaf in early spring. Place it on a piece of graph paper and trace around it. Take a leaf from the same plant in August. Trace around it and compare the two. See how much the leaves of the gourd plant have grown throughout the summer.

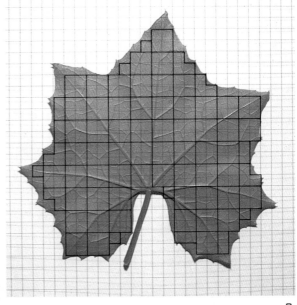

◀ **A gourd plant tendril coiling around a rope (photos 1-4).**

The tendrils of this loofah gourd plant grow quickly. A new coil forms every two hours. The coiled tendril works like a spring (arrow) in supporting the plant along the rope.

▶ **A gourd plant along a support fence.**

Like cucumber and pumpkin plants, gourd plants develop vinelike stems with large leaves. They spread out quickly in many directions. If gourd plants are allowed to grow along the ground, they soon may take up all the space in a garden. So gourd plants, like pea plants and bean plants, are often made to grow upwards. A support pole or trellis is placed nearby so the long, slender tendrils of the gourd plant can coil around it. As the plant grows upward, its leaves are exposed to more and more sunlight.

But sometimes the fruits (gourds) that the plant produces are so large and heavy that it is not practical to use support poles for the plants.

▶ **How a tendril works.**

Tendrils work like springs. Once they have attached themselves to a supporting pole, they become tough and strong and it is not easy to remove them. They give the gourd plant support, and some flexibility, as it grows upwards.

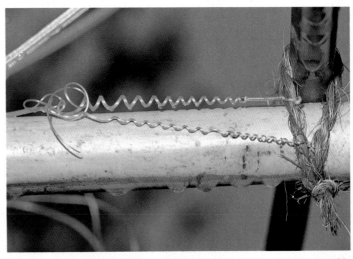

▼ Ants attracted to a gourd plant's nectaries.

The loofah gourd has nectaries near its flower buds, as well as on the back of its leaves. Ants and other insects like to eat the sweet nectar.

◀ **How a gourd stem grows.**

This gourd vine was marked in segments of the same length to show where plant growth takes place (left photo). After the plant had grown for a few days, the right photo was taken. It shows that most of the growth occurred at the tip of the stem. This is true for most kinds of plants.

The stem (vine) of the gourd plant supports the leaves and holds them up to the sun. The leaves must absorb sunlight in order for photosynthesis to take place. Food that is produced in the leaves is carried to other parts of the plant by tubelike veins in the stem, called vascular bundles. The stem also transports water and nutrients absorbed from the soil by the plant's roots to other parts of the plant.

Soon, flower buds begin to form along the gourd vine. Later, fruits (gourds) will develop from some of the flowers.

▶ **Flower buds that do not bloom.**

When a gourd plant is young, the flower buds do not bloom. They turn brown and fall from the plant (right photo). The photo at the left shows a female flower bud (black arrow) and a male flower bud (yellow arrow).

◄ **Male gourd flowers.**

The male flower buds on gourd plants grow in groups, or clusters. There may be as many as ten buds in each cluster. The lowest bud in the cluster is the oldest and blooms first, then the next lowest, and so on.

structure of a male flower

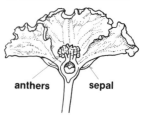

anthers   sepal

When the gourd plant is young, tiny flower buds form, but they soon wither and die without blooming. But by early summer, the plant has developed a long vine and many large leaves, and it is ready to bloom.

Flowers are important to plants because they produce seeds from which new plants grow. Most gourd plants have large, showy, yellow flowers. The brightly colored petals attract honeybees and other insects.

Both male and female flowers are found on the same gourd plant. Generally, the male flowers bloom first.

► **A male flower blooms (photos 1-4).**

It takes about twelve hours for the flower of this loofah gourd to open. It starts to unfold in mid-afternoon and is completely open by early the next morning. If the weather is cold, it may take longer for the flowers to open.

▲ A honeybee flying to a gourd flower.

The fully opened flowers of the loofah gourd measure two to four inches across. Bees cannot see red, but they are attracted to bright yellow flowers like these.

▼ A female gourd flower blooming.

Inside the petals of the female flower is a long, thick tube, the pistil. The tip of the pistil is called the stigma. The long, narrow neck is the style. At the base of the pistil is the ovary, where the eggs are located, from which new seeds will form.

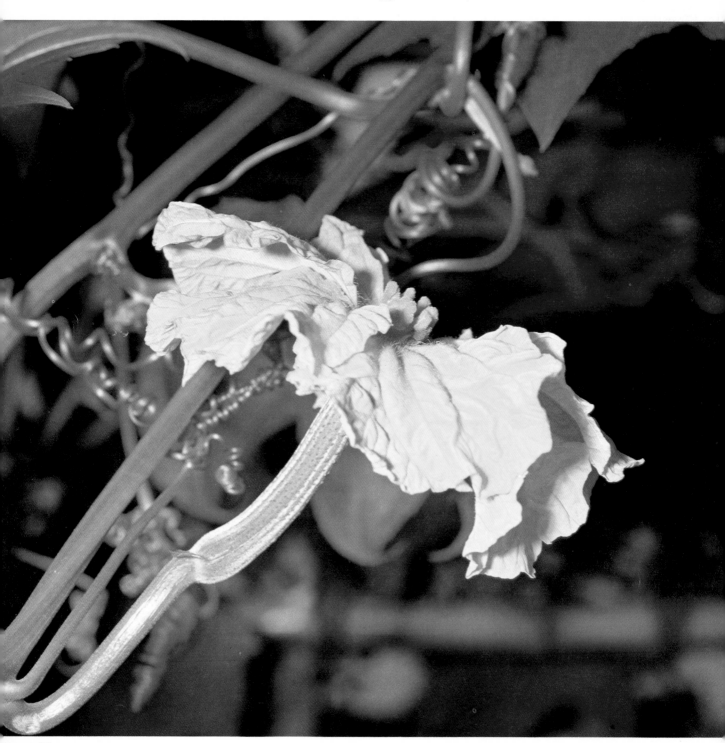

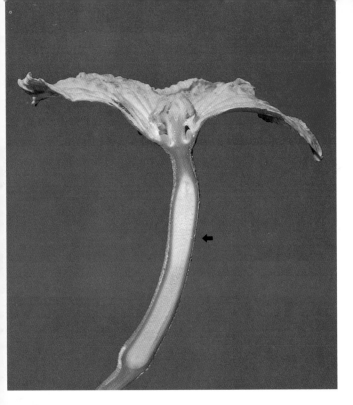

● **A cross-section of a female flower (left) and a male flower (above).**

The female flower has an ovary (arrow) where the fruit, or gourd, will form. Male gourd flowers have stamens with yellow tips, called anthers. The anthers contain tiny, dustlike grains of pollen that enclose the male sperm cells.

After the first male flowers have bloomed, the female flowers begin to bloom, as well. The female flowers are not arranged in clusters, as the males are. There is only one female flower on each flower stalk, or peduncle. Both the male and female flowers of the loofah gourd plant remain open only for a day. But new flowers will continue to bloom on the plant day after day, throughout the summer.

▶ The weight of this gourd flower has snapped the flower stalk in two (left photo), and a camel cricket is feeding on a gourd flower (right photo).

● **Insects gathering pollen and nectar from gourd flowers (photos 1-4).**

(1) A skipper butterfly collecting nectar. Grains of pollen are clinging to the insect's long, strawlike mouth, called a proboscis. (2) A dronefly on a male gourd flower. (3) A honeybee using its long proboscis to drink plant nectar. (4) Honeybees gathered on a female gourd flower.

In order for new gourd plants to form, a process called pollination must take place. That means a pollen grain from an anther must touch the stigma of a pistil. Sometimes pollen is carried from one plant to another by insects or the wind. This is called cross-pollination. Most gourd plants are cross-pollinated. But they also have the ability to self-pollinate because they have both male and female flowers on the same plant.

Often, insects help pollination to take place. Bees, butterflies, and other insects are attracted to the pollen and nectar in flowers. Nectar is secreted by plants especially to attract insects. It has lots of sugar in it and gives insects the energy they need to move around. Plant pollen is rich in protein. It has the same food value for insects that meat and eggs have for people.

When an insect flies to a male flower, searching for pollen and nectar, grains of pollen stick to its body. When the insect lands on a female flower, some pollen brushes off onto the stigma, the tip of the pistil, and pollination has taken place.

The pollen grain absorbs water and sugar from the stigma and begins to swell up. Soon, it sends a narrow tube down the long neck, or style, of the pistil. When it reaches the ovary where the egg cells are, a sperm from the pollen joins with an egg, fertilizing it. After the egg has been fertilized, a plant seed begins to grow.

◀ Gourd fruits begin to grow.

Within a few days, gourds begin to grow from the pollinated flowers. The flower itself will soon wither, but the gourd will grow quickly.

▶ Mature gourds.

When loofah gourd plants grow along support poles, the gourds grow downward, in a vertical position. The plant's strong vine can hold up even the heavier, mature gourds.

Once the eggs are fertilized by the sperm, seeds begin to form. As the seeds develop, the ovary swells to protect them. The swollen ovary becomes the fruit of the gourd plant, or the gourd itself. A thick, outer skin, or rind, forms over the gourd, giving the seeds even more protection. The loofah gourd rind is green throughout most of the summer. It turns yellow or brownish in the fall as it ripens. Ridges form on the rind of this species of loofah gourd.

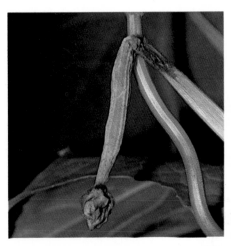

◀ A mature gourd.

Although the flower petals have fallen off, you can still see the long, green style at the tip of this gourd.

▶ An unpollinated flower.

Not every female flower on the gourd plant produces fruit. Some are never pollinated. They wither and die without producing fruit.

▲ Hummingbird hawk moths sipping flower nectar.

The helping relationship between gourds and insects continues throughout the summer. Gourd plants continue to flower, providing pollen and nectar, nutritious food for hungry insects. And as butterflies and bees flit from flower to flower, they carry grains of pollen with them, helping to pollinate the female flowers. New gourds form throughout the summer as new flowers are pollinated.

Some insects, particularly moths, search for pollen and nectar in the evening. The bright yellow flowers of the gourd plant can easily be seen at dusk.

▶ Skipper butterflies visiting a male gourd flower.

▼ The growth of a gourd fruit.

The photo compares the sizes of a baby gourd (left), young gourd (middle), and fairly mature gourd (right). Each square on the graph paper represents one-fifth of an inch. Gourds grow in both length and width. The green leaflike structures at the tip of the gourd are the sepals that form the calyx, which enclosed the flower bud.

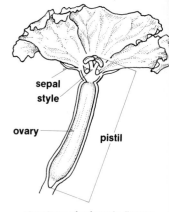

sepal
style
ovary
pistil

structure of a female flower

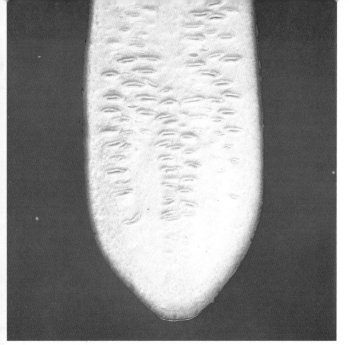

▲ A cross-section of a young loofah gourd.

The flat white seeds of the young gourd are surrounded by a tender white pulp.

▲ A cross-section of a mature gourd.

The seeds have hardened and turned black. The pulp has become woody and tough.

If you were to cut open a young loofah gourd, you would find that the rind is still tender. Inside, there are many white seeds surrounded by a fluffy white pulp, which is the ovary. People in Japan sometimes eat young loofah gourds, while the fruit is still tender.

It is much harder to cut through the rind of a mature gourd. The skin is much tougher. The pulp inside has become tough and woody. In a mature gourd, the seed coats have hardened and turned black.

▶ The growth of a seed.

Once an egg is fertilized, a seed begins to form (left photo). Seed leaves form inside the seed coat. The right photo shows a cross-section of a mature seed.

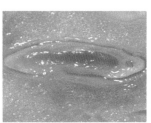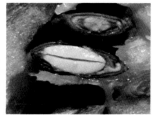

▲ A tendril which has begun to dry out and turn brown in the fall.

◀ A fruit cricket chirping on a dried leaf.

In autumn, as the leaves of the gourd plant dry up and turn brown, crickets and other fall insects begin to appear.

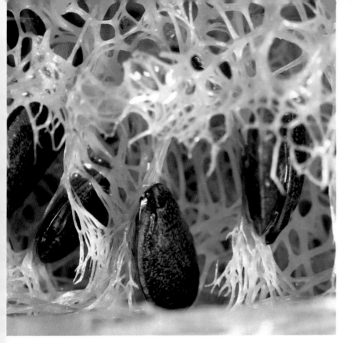

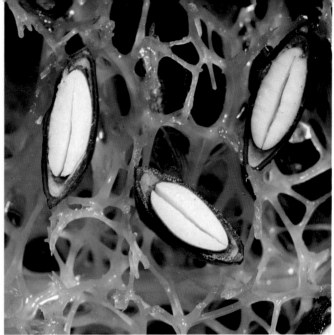

▲ **Mature seeds surrounded by plant fibers.**

This is what the inside of the loofah gourd looks like after the pulp has been washed away. Only the spongelike network of fibers remains, with a few seeds still attached.

▲ **A cross-section of mature seeds.**

Each seed has a pair of seed leaves. These are protected by a tough seed coat. When the seeds are planted in the spring, new gourd plants will grow from them.

By fall, the leaves and vine of the gourd plant have withered and turned brown. The loofah gourds themselves, which had been green all summer, turn yellow or brownish. They may measure several feet long.

Once the ripened gourds are picked, they are soaked in water for a few days. Then the rind can be easily peeled off and the pulp washed away. Only a lacy network of strong plant fibers remains, which looks very much like a sponge. In fact, the loofah gourd is also called the dustcloth or bath sponge gourd. Its tough fibers are ideal for scrubbing pots and pans and are used as bath sponges, particularly in Europe and Japan.

◀ **Dried gourds.**

If gourd fruits are left on the plant after they have ripened, they dry out and become lightweight. By the time the gourd rind turns brown, the tip of the gourd breaks open.

▶ **Gourd seeds.**

The tip of the dried-out loofah gourd breaks open and the seeds spill out. They are scattered in many directions by the wind.

Mature loofah gourds that are left on the vine begin to dry out in the strong winds and autumn sunshine. One day, the tip of the loofah gourd breaks off and the seeds fall out. Each seed contains a tiny plant embryo with nutrients stored in the seed leaves, all that is necessary to form a new plant.

The wind scatters the seeds in many directions. Some, but not all, will be eaten by animals. A few will survive the winter cold, protected inside their hardened seed coats. In spring, if the growing conditions are right—good soil, enough rainfall, and plenty of sunlight—the seeds will sprout, or germinate. New gourd plants will begin to grow.

Let's Find Out More About Gourds.

● **Bottle gourds (photos 1-5).** (1) Ripe bottle gourds. (2) A male flower. (3) A female flower. (4) Young bottle gourds. (5) A cross-section of a young bottle gourd with seeds.

There are many different kinds of gourds in the world, almost a thousand species in all. Some are raised by commercial growers. Many others grow in the wild. The gourds pictured above are called bottle gourds. Long ago, when people didn't have pottery or dishes, they used bottle gourds as drinking cups and food containers. Today, most gourds are grown for ornamental purposes. There are various sizes, shapes, and colors of gourds.

Some flowers, like the one below, contain both male and female parts. This is not true of gourd plants, which have separate male and female flowers.

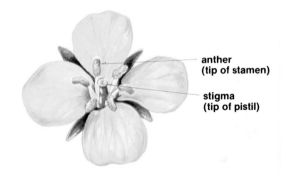

anther
(tip of stamen)

stigma
(tip of pistil)

Cucumbers

Cucumbers came from India originally. Like gourds, cucumbers have large, showy, yellow flowers. They have both male and female flowers on the same plant.

A female cucumber flower.   A male cucumber flower.

Pumpkins

Scientists believe pumpkins originated in South America. There are many varieties of pumpkins. The thick vines and large leaves of the pumpkin plant are similar to those of the loofah gourd plant. Pumpkins are widely grown for food.

A female pumpkin flower.   A male pumpkin flower.

Watermelons

Watermelons are from Africa. The fruit of the watermelon is usually red, but it may be pink or yellow. Watermelons are sweet and juicy. One melon may have as many as 500 seeds in it.

A female watermelon flower.   A male watermelon flower.

The Gourd Family

Cucumbers, squash, pumpkins, melons, and watermelons are all members of the gourd family. They are sometimes called edible gourds. The Latin name for the gourd family is Cucurbitaceae. In the photos on this page, you can see similarities in the fruits and flowers of different members of the gourd family.

A female melon cucumber flower.   The fruits of a snake gourd plant.

GLOSSARY

embryo—the early stages of development of a plant or other organism. Plant embryos form inside the seed coat. (p. 28)

fertilized—when an egg and a sperm unite, making it possible for a new organism to form. (p. 19)

germinate—when a seed begins to grow, or sprout. (p. 28)

nectar—a sweet liquid secreted by plants especially to attract insects. (pp. 6, 12, 19)

nectaries—the glands in plants that secrete nectar. (pp. 6, 12)

photosynthesis—the complex process by which green plants make food, with the help of chlorophyll, a substance found in the plants' leaves, and energy from sunlight. (pp. 6, 13)

pollination—the process in which pollen is transferred from an anther to the tip, or stigma, of a pistil. (pp. 19, 22)

proboscis—a tubelike mouth used for sucking liquids or body fluids. (p. 19)

seed leaves—the first leaves to form on a plant. Seed leaves are part of the embryo. They form inside the seed coat and contain stored nutrients. (pp. 4, 6, 28)

vascular bundles—tubelike veins in a plant stem or leaf in which food and water are carried from one part of the plant to another. (p. 13)